BOOK · 3

THE HO[illegible]E AT GRASS

THE MANUAL OF STABLE MANAGEMENT

The Kenilworth Press Limited

COMPILED BY

Pat Smallwood FBHS

THE ADVISORY PANEL INCLUDED

Stewart Hastie MRCVS

Jeremy Houghton-Brown B Phil Ed

Clive Scott

Tessa Martin-Bird FBHS

Barbara Slane-Fleming FBHS

Helen Webber FBHS

SERIES EDITOR

Jane Kidd

Reprinted 1990
Reprinted in enlarged format 1992, 1993, 1995

British Library Cataloguing in Publication Data
A catalogue record for this book is available from the British Library

ISBN 1-872082-27-0

Produced for the British Horse Society by
The Kenilworth Press Ltd
Addington, Buckingham, MK18 2JR

Printed and bound in Great Britain by
Hollen Street Press Ltd, Berwick upon Tweed

CONTENTS

Introduction

The aim of this series is to provide a reliable source of information and advice on all practical aspects of horse and stable management. Throughout the series emphasis is placed on the adoption of correct and safe procedures for the welfare of all who come into contact with horses, as well for the animals themselves.

The books have been compiled by a panel of experts, each drawing on considerable experience and contributing specialised knowledge on his or her chosen subject.

The other titles in the series are:

Book 1, The Horse – Conformation; Action; Psychology of the Horse; Teeth and Ageing; Breeds; Breeding; Identification; Buying and Selling; Glossary of Terms.

Book 2, Care of the Horse – Handling the Horse; Stable Vices and Problem Behaviour; Grooming; Bedding; Clipping, Trimming, Pulling and Plaiting; Recognising Good Health and Caring for the Sick Horse; Internal Parasites; Shoeing.

Book 4, Saddlery – Saddles; Bridles; Other Saddlery; Bits; Boots and Bandages; Clothing; Care and Cleaning of Leather; Saddling and Unsaddling.

Book 5, Specialist Care of the Competition Horse Dressage Horse; Driving Horse; Show Jumper; Event Horse; Long-Distance Horse; Hunter; Show Horse or Pony; Point-to-Pointer; Polo Pony; Types of Transportation; Travelling.

Book 6, The Stable Yard – Construction; Riding Schools; Organising and Running a Yard; The Buying of Fodder and Bedding; The Law.

Book 7, Watering and Feeding – Watering; Natural Feeding; The Digestive System; Principles of Feeding; Foodstuffs; Rations; Problem Eaters; The Feed Shed, Storage and Bulk Purchasing.

NOTE: Throughout the book the term 'horses' is used and it will often include ponies.

CHAPTER 1

Grassland Management

Constant use of the land for grazing by horses will always present problems.

Horses are very selective in their grazing habits. Unlike ruminants, who spend much of their day resting and chewing their cud, horses graze more or less continuously – except for short periods of rest or when sheltering from hot sun and flies. Fields grazed only by horses, even with a low number on a large acreage, develop 'roughs' where they dung (also called 'camp areas') and 'lawns' where they graze. Only if very short of food will horses graze on these 'roughs', and unless good management is practised, the 'roughs' increase in size and so reduce the grazing area. They also become reservoirs of infection for internal parasites and a home for infective larvae. In wet weather, the latter are able to migrate to the 'lawns' with a corresponding greater risk of infection to the grazing horse. Thus fields grazed entirely by horses are likely to deteriorate, unless given knowledgeable and experienced management.

The objectives of good management are:

1. To provide grazing over a long period of time by the establishment and/or maintenance of a dense vigorous sward of suitable variety.
2. To minimise infection from worm larvae, although in comparison to anthelmintic strategy this plays a minor role.

3. To provide suitable working areas: i.e. for jumping and schooling.
4. To maintain the land as a visual amenity, thus avoiding complaints from neighbours and possible action by local authorities. Well-kept fields and fencing are a good advertisement for any commercial establishment, and a source of satisfaction to the private owner.

Management must be geared to provide conditions which encourage the most productive and palatable grasses and discourage weeds and less productive species. Palatability tends to be a reflection of good management rather than of species.

The following procedures are recommended:

Soil Analysis

This should be carried out every four or five years, and when moving into a new establishment, particularly if the grassland appears to have been neglected.

The Ministry of Agriculture through their local Advisory Service Office (ADAS) advise on local seed and fertiliser merchants qualified to carry out soil analysis. Results are in accordance with the ADAS index. In fields grazed only by horses, soil samples should be taken both from the 'lawn' and the 'rough' areas. The latter are likely to be high in potash as a result of horses urinating in the area. According to the results of the analysis, lime, phosphate and potash can then be applied.

Beware of expensive 'trace element' fertilisers and sprays. If horses are found to be deficient in particular trace elements, it is usually more effective to feed or inject these, rather than to make up the deficiency in the soil. Breeding establishments may need to take specialist advice.

Drainage

Signs of poor drainage are:

☐ Surface water.

- Plants such as rushes, tussock grass, buttercups and couch grass.

Any drainage scheme is expensive and may be beyond the finances of a small establishment, but if the field is part of an agricultural holding it may be possible to obtain a grant.

Before embarking on a drainage scheme, make sure that all boundary ditches and outlets are cleared, as these can be one of the main causes of surface water in a field. Drainage problems can also result from damage to soil structure, by the use of heavy machinery when the ground is wet, or by 'poaching' from animals left out in the winter. If damage to soil structure is suspected, advice should be sought from ADAS. Improvement can usually be achieved by subsoiling and pan bursting. Such work has to be carried out by a specialist firm.

If a drainage system has to be installed, its design depends on soil structure, the lie of the land, and other natural features. In areas of clay soil, underground channels – preferably in a herringbone design and connected to piped major drains – can be successful. It is possible to install drainage pipes by machines without disturbing the top surface. In all cases, professional advice from a reputable firm should be sought.

Harrowing

Before allowing tractors on to a field, the surface should be sufficiently firm to take the weight without damaging soil structure.

Harrowing should begin as soon as the land is dry enough. Depending on the season and the soil, this may range from late February to early April. A spiked or pitch pole type harrow is the most efficient, as it removes the dead grass and moss, aerates the soil, and encourages new growth. Chain harrowing to spread the dung and tidy the paddock may be carried out through the summer, when paddocks are rested, and preferably during hot, dry weather. In a dry season, the hot sun helps to dessicate the manure and kill off worm eggs and larvae. In a wet season, both eggs and larvae tend to thrive, so there is a danger in spreading the droppings over a wide area. They could increase the area of tainted pasture, especially in fields which do not drain well.

Rolling

Rolling is beneficial:

- ☐ If a severely poached field is to be used for riding. Harrowing followed by rolling will provide a level surface. To achieve a good result the soil must be dry enough not to stick, but not so dry that the lumps of earth will not break up.
- ☐ If the field is to be cut for hay. Rolling can reduce the risk of damage by hay-making machinery when it is travelling over a rough surface.
- ☐ In early spring, when it can encourage early growth by raising soil temperature.

A light-ribbed 'Cambridge' roller is most suitable. With certain soils – particularly clay and heavy loam – there is a risk that rolling may damage the soil structure and assist in the formation of an impervious pan. This inhibits growth and causes drainage problems. Such areas should not be rolled unless this is absolutely essential.

Fertilisers

In areas where soil fertility is good (usually a sign of careful management over many years), the application of fertiliser may not be essential. Lush grass is not required for horses, and unless other stock is available to graze off the first growth, the application of fertiliser can be unnecessary and indeed unwise. Ponies are quick to gorge on new grass, with resultant digestive upsets and risk of laminitis. Establishments with a limited acreage find that the application of a suitable fertiliser is essential if grass is to be encouraged and weeds kept in check.

The four major elements necessary for healthy plant growth are lime, phosphate, potash and nitrogen. Magnesium is included in the soil analysis, but appears to be of no importance to the health or growth of a horse. Nitrogen is not included, as amounts vary according to the season and rainfall.

Following the soil analysis, local fertiliser firms who know about the soil features in the district can be consulted. Representatives may need to be reminded that horses have different grazing requirements from milking cows, beef cattle, and sheep. A steady growth over a longer period is needed, not a lush growth of grass.

Lime. Levels are measured on a 'PH' scale, 7.0 being neutral. A 'PH' level of 6.5 is suitable for grass – below this level the soil is acid, leading to the growth of poorer grass – above PH7 the soil is alkaline and the large amount of calcium results in the locking up of certain minerals. A high PH (excess lime) can seriously affect the growth of young stock. A level above PH7 is inevitable on chalk soil.

A deficiency of lime is usually adjusted by a dressing of limestone or chalk, both of which contain the calcium which is so important for healthy bone growth in the young horse. This dressing can be applied at any time of the year, but preferably in autumn or spring and on a calm day to avoid drift loss by the wind. In the past, basic slag was used, but it is now difficult to obtain.

Phosphorus. (P_2O_5 is phosphate.) The Ministry of Agriculture have drawn up an ADAS index from 0 – 8. For horses, a suitable index should be 1 – 2. This allows for the phosphate level to be below the calcium level. This correct balance is essential for good bone development in young stock and general health. A low phosphate level, i.e. index 0, hinders grass growth.

Rock phosphate. This should not be applied unless the soil is very acid (below PH5), but experiments are continuing to make rock phosphate more soluble and thus more readily available to plants. Super-phosphate can cause sudden bone failure and should not be used.

Potassium. (K_2O_1 is potash.) Potash is necessary for plant growth. An index level of ADAS 1 – 2 is suitable for grazing land. If hay is to be cut, then a dressing of potash after the hay has been taken corrects any loss. In general, if land is grazed by horses, potash levels once established remain constant. Little is known about the ingestion of this mineral by horses.

Nitrogen (N). Nitrogen is the key to grass growth, although success also depends on the correct balance of the lime, phosphate and potash. Nitrogen is leaked out of the soil by rain, so it is best applied during the spring and summer months. Swards containing large amounts of clover manufacture their own organic nitrogen from bacteria on the clover root. It can make them self-supporting in nitrogen, but the amount of clover needed to achieve this is too much for horses. It can lead to digestive upsets and laminitis.

Too much nitrogen on grazing land, with resultant over-fast growth, can produce unsuitable grasses and a lack of nutrients. This can affect the bone growth of young horses.

Application of Fertilisers

These are best applied in the spring when the field is not being grazed. If the grasses are to take full advantage of the treatment, a period of rest is advisable. Should this not be possible, the fields should be safe to graze once a liquid fertiliser has been washed in by a light shower or heavy dew, or the dry granules are not visible on the surface.

To save expense, a suitable mix of required fertilisers in either granulated or liquid form may be put on in one operation as a compound. Since grass growth for horses is required over a long period, it may be advisable to apply a small amount of nitrogen in the spring and some more later when the fields or paddocks are being rested.

Some people object to the continual use of chemical fertilisers. Owners of breeding stock often prefer to use organic fertilisers (such as FYM or seaweed). These are slower-acting and more expensive but provide minerals which may otherwise be lacking. Adult horses, other than brood mares, are unlikely to be affected by the use of chemical fertilisers. There is normally a second natural growth of grass in September, and fertiliser applied in mid-August promotes this.

Manures

Farmyard Manure (FYM). If available, FYM is a complete fertiliser. It also sweetens those parts of a field which have been soured and made 'horse sick'. It provides material to improve soil texture, corrects any tendency of the soil to be alkaline, and supplies the main plant nutrients of phosphate, potash and nitrogen.

It is best used on fields intended for hay and should be applied in the autumn when the weather and earthworm population can work it into the soil. If applied on grazing fields it should be stored for at least six weeks before use and the field should not be grazed for six weeks after spreading, to ensure that any infection is eliminated.

Unless linked to a farm or small animal enterprise, the average horse establishment is unlikely to have any opportunity of obtaining FYM.

It is sometimes possible to obtain cattle or pig slurry and also human sewage sludge from council sewage farms. The latter is treated and sterilised. All of these make excellent fertilisers, but are strong and therefore they should only be used in small quantities and not on grazing fields. They are suitable for hay fields, encouraging strong growth.

Once spread, the strong smell of any FYM or slurry can linger for days, which is a consideration for establishments dealing with the general public.

Horse Manure. This is not as valuable as FYM and is generally inadvisable for fields grazed by horses, as it is likely to increase the area of tainted grass. There is also some risk of worm infection, as it requires considerable heat over several weeks to kill the eggs and worm larvae. Manure must be left to heat and rot before applying.

Early Grazing
Grass starts to grow as soon as (a) soil temperature rises and (b) there is sufficient rainfall. If the sward is to profit from the treatment already outlined, it must not be grazed until the grasses are well established: i.e. approximately 10cms (4ins) tall. Earlier grazing is to the detriment of the better-quality grasses, which need to develop strong roots and invariably grow more slowly than the poorer grasses and weeds.

Rotation
Where there is sufficient acreage, a system of rotational grazing should be practised. This involves dividing large fields into smaller paddocks. An area of 3 acres is a useful grazing size. The divisions can be achieved either with permanent fencing or by the use of electric fencing (see *Fences*, page 33). In early summer – i.e. May or June – the stocking rate could be as high as ten horses or fifteen ponies per paddock for periods from ten to fourteen days. The paddock may then need top-cutting (see below), and should be rested for three weeks before being grazed again. In July and August grass growth slows down and the paddocks take longer to recover. If more growth is required, they may benefit from some nitrogen. Growth accelerates again in September, but the first frosts check it until the following spring. In June, the grass which is surplus to grazing requirements may be cut for hay. An alternative is to cut the grass and use it for fresh green feed for stabled horses. If possible, the paddocks chosen to be cut for hay or green feed should be rotated each year.

Top-Cutting
From June onwards, it may be necessary to mow over lightly or top-cut all paddocks. If horses stay in the field and if the weather is fine, much of this taller grass may be eaten as the sun helps to sweeten it. Any heavy patches of grass which remain should be forked up and burned. If left, they check the re-growth of the underneath grasses.

This top-cutting is necessary to prevent weeds and uneaten coarse grasses from seeding and to encourage growth from the better grasses.

When you are top-cutting, a careful watch should be kept for ragwort. It is sensible to have the paddocks searched before cutting is started. The ragwort plants should be pulled up and burned. This task is most easily carried out after rain, when the ground is soft.

Over-Grazing

Fields which are over-grazed, and the sward not allowed time to recover and establish new growth, are bound to deteriorate. Such treatment can result only in the encouragement of weeds and coarse unpalatable grasses; an increase in worm infestation; and the loss of what could be valuable grazing. Where the amount of land is limited, it is sensible to try to obtain alternative grazing, and certainly to avoid grazing any damp fields between October and May.

It should also be noted that the sward deteriorates if under-grazed and the grass allowed to seed.

Cattle as Alternative Grazers

Cattle make excellent followers-on for paddocks grazed by horses. They eat the coarse grasses rejected by horses, and ingest and kill the worm larvae deposited on the soil from the horse's dung. Sheep and goats can also help to free grass from worm larvae, but may require special fencing. Sheep in particular prefer short sweet grass, similar to that enjoyed by horses. To buy cattle or sheep for the above purpose can be a considerable capital expense with no guarantee of a cash profit at the end of the summer. As an alternative, it may be possible to arrange for a neighbouring farmer to graze his cattle in return for hay or help with upkeep and field cultivation.

Picking up of Droppings

If time and labour are available, the daily picking up of droppings in small paddocks is worthwhile. It decreases the worm infection, improves the appearance of the field, and encourages the growth of more palatable grass. A vacuum machine, which collects droppings and can be attached to a tractor, is now on the market.

Poaching

If animals are kept out during the winter months considerable damage can be done to a sward. Low-lying or clay soil areas are likely to be worse affected. Damage is caused by deep hoof prints in the sward, which in a wet season often remain full of water, adversely affecting the root system of the grasses. Exposure to frost causes further damage. If wintering out animals, it is sensible to use only one or two paddocks, preferably the best drained. Shelter, water and access must also be considered. In low-lying areas, in a wet season it is sensible for fields to be shut up in October and not grazed again until April or May. The decision to close the field depends on the season. Well-drained and/or hill fields may be safely grazed through the winter.

Stocking Rate

This varies according to the following factors:

- ☐ Soil type and fertility.
- ☐ Length of grazing season.
- ☐ Rainfall.
- ☐ Period of daily grazing.
- ☐ Area of land available and whether divided into separate paddocks.
- ☐ Parasite control.
- ☐ Quality of management.

Given good soil fertility, an approximate stocking rate throughout the years would be one horse or two ponies per acre, or two horses or five ponies per hectare. From October until the following May, supplementary feeding will be necessary. On land with poor fertility, this rate of stocking could be halved: one horse or two ponies per two acres. Supplementary feeding may be necessary from July onwards, particularly if the animals are working.

If rotational grazing is practised from May to October, three 3-acre paddocks with periods of rest should accommodate ten horses or fifteen ponies. The period of daily grazing affects the consumption of grass. Many private and riding school animals stand in during the day. Others, if working, may need to have their grazing time limited so that they will not become overweight.

IMPROVEMENT OF GRASSLAND

Reduction of 'Camp' Areas or 'Roughs'

These are parts of the field which the horses contaminate by passing dung and urinating. The grass becomes coarse, horses refuse to graze it, and the sward rapidly deteriorates. The situation is worse on poorly drained soil.

Procedure:

- ☐ Daily removal of droppings.
- ☐ Alternate grazing with cattle. Cattle eat down the coarse grass and also ingest and kill the worm larvae and eggs.
- ☐ Regular top-cutting and resting of the field. This allows the grasses and plants to establish a healthy root system, which leads to increased growth.
- ☐ Use of farmyard manure.

Control of Weeds

- ☐ Good management encourages the better grasses, and discourages weeds: buttercups in particular.
- ☐ Some weeds, such as nettles, thistles and bracken, can be controlled or eradicated by regular cutting: i.e. six times in the growing season. This method may have to be repeated for a second year before success is achieved.
- ☐ Clumps of weeds, i.e. nettles, thistles and docks, can be treated by using a suitable herbicide applied directly on the clumps by a knapsack sprayer.

☐ If the weeds are widespread, total spraying of the field may be necessary. A weed-killer can often be combined with a fertiliser to give a boost to the fine-leaved grasses. It should be remembered that some selective weed-killers are effective against all broad-leaved species, and thus many other beneficial and appetising herbs and grasses will be destroyed. More sophisticated weed-killers are now available which may avoid this problem. There is new legislation covering spraying weed-killers. This should be checked before any spraying is done.

Weed-killers should be applied in late spring when the plants are growing vigorously. In some cases, a second application will be necessary later in the summer when regrowth has occurred. Spraying should take place in calm weather to avoid drift and when rain is not expected for twenty-four hours. Some sprays are toxic to stock, so fields should not be grazed for some time after spraying. If poisonous plants are present – i.e. ragwort – they must have withered and crumbled before stock can return to the field.

Agricultural merchants can usually be trusted to give some advice as to the necessary type and quantity of weed-killer to use. For some species – e.g. docks and bracken – this has to be applied later in the summer. They will also advise as to which weed-killers are toxic to stock. It is important to remember that horses are much more sensitive to weed-killers than other stock.

Direct Drilling

Small areas of badly poached ground should be harrowed or, if necessary, raked over by hand. Suitable horse paddock mixture should then be broadcast, and the area should be rolled. For small areas a grass 'hand seeder' (on sale in specialist shops) can be used to introduce herbs into a pasture which is deficient in them. It is then advisable to treat a narrow strip of the field.

Whole fields can be improved by direct drilling or slot

seeding. This work is usually undertaken by contractors. Advice should be taken as to whether (a) the soil is suitable and (b) the process is likely to succeed. The best times of year are spring or late summer, when the soil is warm and there is sufficient rain to germinate the seed. It may be advisable to kill off the poor-quality turf and weeds with herbicide. This will not affect the new seed and will ensure that the new grass has a good start and is not smothered by already established seeds. A special drill is used which cuts into the ground, making channels into which the grass seed and fertiliser are directed. The field is then rolled.

This process does not disturb the top layer of soil and encourages the rapid establishment of a dense sward, which can be grazed by horses at a much earlier date. It is of particular value in fields where stones and flints are close to the surface. If such fields are ploughed, the surface can be unsuitable for riding and schooling horses for several years.

Ploughing and Re-Seeding

It is now considered more sensible to improve pasture by good management and perhaps by direct drilling, rather than by ploughing up and re-seeding. Ploughing should only be resorted to in urgent cases where the sward appears to have gone beyond reclaim. Advice should be sought from the local ADAS office.

Ploughing and re-seeding are usually undertaken by local contractors in late summer. The old poor-quality, weed-infested turf can be killed off by herbicide. The field is then ploughed, and then scarified to make a suitable seed bed. Seed and fertiliser are drilled together and the field is rolled. The following summer, the field is best grazed by sheep to consolidate and encourage a dense sward. In the first year it should not be ridden on, but a hay crop may be taken. It can be grazed by horses in late summer, but in a wet season care should be taken to ensure that it does not get poached. If the new grasses are to thrive it will require careful management. There will be competition from indigenous plants, which may have been germinated with the ploughing.

Special seed mixtures are now available for horse pasture. Advice should be taken as to whether they are suitable for the particular area. They contain appropriate quantities of grasses palatable to horses and some clover and selected herbs.

GRASSES

The requirements for horse pasture are prostrate growing grasses with a good bottom growth and the ability to produce a dense sward. The following grasses are of value:

- ☐ Perennial rye grass. There are several varieties, which are all of value.
- ☐ Smooth meadow grass, known in the USA as 'Kentucky blue grass'.
- ☐ Creeping red fescue.
- ☐ Chewing fescue.
- ☐ Sheep's fescue.
- ☐ Tall fescue.
- ☐ White clover in small quantities.
- ☐ Timothy, cocksfoot and the tall early rye grass (Lolium SPP) are all suitable for hay, but do not make good grazing for horses or a hard-wearing sward.

Some grasses, which on fertile soil would be classed as weeds, and undesirable, have advantage where the soil is poor in natural fertility, e.g. hill areas. In such areas, some of the better grasses will not flourish. Crested dog's tail, sweet vernal and common bent grass all flourish on poor soil, but they only provide third-rate pasture (often suitable for ponies).

Comprehensive research is being carried out on the development of new strains of grasses, which may in time supersede the strains recommended above.

HERBS

Horses relish herbs, which provide nutrients often lacking in more shallow-rooted species. The following are of value:

- ☐ Narrow-leaved plantain.
- ☐ Yarrow.
- ☐ Dandelion.
- ☐ Chicory.
- ☐ Comfrey.
- ☐ Ribwort.
- ☐ Burnet.

HAYMAKING

Aspects of haymaking which need consideration are:

- ☐ The sward should be of a reasonable quality. Fields infested with weeds and growing inferior grasses with only a small proportion of good grasses are not suitable.

- ☐ When the fields are shut off for hay, sufficient grazing should be available. Depending on soil fertility, sward and rainfall, hay fields will not be ready for grazing for at least four to six weeks after cutting.

- ☐ It is not economical to buy haymaking machinery for small acreages, so a local contractor or neighbouring farmer should be hired to do the work. Such arrangements often result in small acreages being left until last, which will mean the risk of deteriorating weather, a crop of poor quality, and grass past its best.

- ☐ Fertilisers cost money, but they are necessary if the crop is to be worth harvesting. Contractors charge per acre for cutting and turning. A poor crop of grass can often be just as costly to make as a heavy crop, although baling costs will be less.

The advantages of making your own hay are:

- ☐ The hay (subject to the factors mentioned above) will be of good quality and costs less to make and store than buying from a corn merchant.
- ☐ The fields should benefit from being cut for hay; good grasses will be encouraged and weeds and inferior grass cut before being allowed to seed. The sward if well harrowed improves by not being continuously trodden.
- ☐ In May and June, establishments with ample grazing have an over-abundance of grass. Unless this is (a) set aside for hay, (b) cut as green fodder for stabled horses, or (c) efficiently grazed, the sward deteriorates. Grasses are likely to be swamped by stronger-growing grasses and weeds. Cattle can be brought in to eat off this grass. This is an expensive capital outlay and not necessarily profitable.
- ☐ Shutting up the field helps to ease the worm burden, which is likely to increase if fields are continually grazed by horses.

The disadvantage is:

- ☐ It is very likely that horses and ponies will dislike and will not eat hay that has been made from pastures that have been grazed only by horses.

Preparation of Hay Fields

Meadow hay is made from fields of permanent grass specially 'shut up' (not grazed). Low-lying fields should not be grazed during the winter, in order to avoid poaching of the ground. In early spring, fields should be harrowed and fertiliser should be spread; rolling is not usually necessary, and can be detrimental by increasing panning. The grass can then be left to grow. Before cutting it must be checked for ragwort.

On dry, well-drained fields it is possible to allow grazing during the winter and not to 'shut up' until mid-April. Harrowing and fertilising can then be carried out. Though the resultant crop will not be as heavy and will be late growing, it can provide a useful supply of hay.

Seed hay is made from fields specially sown down. They may not have been grazed at all; the turf may not be established; and grazing – particularly if the ground is soft – would seriously damage the sward. The crop is much heavier than that from permanent pasture, and may take longer to 'make'. It should be of superior quality and free from weeds.

Making the Hay

Hay is made by cutting grass: preferably just before flowering, but if the weather is uncertain it may have to be delayed. The grass is left lying in rows. After a few hours of hot sun the top surface begins to wilt. The grass is scattered by a machine, a process known as 'teddering', which scatters the rows of grass and allows wind and sun to penetrate and dry the crop. A heavy crop will need teddering at least twice; after rain it may need to be done several times. When the hay feels dry and crisp it can be 'rowed up' (two rows put into one) ready for baling.

The hay should then be baled • If the baling is not completed in a day, the hay must be turned again the following morning so that it will be dry and crisp • If the baler is drawing a sledge, the bales can be left in convenient stands of eight to twelve. This makes for quicker and more convenient carting • If there is no sledge, the bales should be stood up in groups of four • The knots of the baling strings should be at the bottom of the bale nearest to the ground. Hay stacked in this manner 'shoots' any rain • It is usual to cart the bales as soon as possible. However, if the weather is settled, the bales improve by being left out in the field for a few days before carting, so that they can sweat and dry naturally. There will then be no overheating.

The making of good hay is a skilled process. Fine weather is essential. In a wet summer, good hay can be difficult to make and find.

CHAPTER 2

Management of Horses and Ponies at Grass

In most parts of Britain, ponies and cobs can be kept out throughout the year. It is their natural lifestyle, especially if they are part of a herd. If suitably fed and properly looked after, they keep healthy and sufficiently fit for hacking and light to medium work. Most of them can withstand dry cold and freezing temperatures. The worst possible conditions are cold, driving rain, inadequate shelter and a wet, poorly drained field. In such weather conditions, horses benefit from New Zealand rugs.

The better quality animals with lighter coats do less well in winter, and will benefit from being stabled at night. Those turned out by day may wear a New Zealand rug (see *Clothing*, Book 4). If stabling is not available, quality animals need extra concentrate food to maintain body heat. This should maintain their health, but if they are to keep their condition they must be very well looked after.

FIELDS

Appropriate Fields

Though your choice may be limited, before buying or renting a field to be grazed by horses you should first consider the location, the type of land and the keep. The following factors should be considered:

- ☐ The field must be easily accessible in winter, so that the horse can be fed and the water supply checked.
- ☐ Well-managed, old-established ley or meadow field is the best type of grazing. It provides a great variety of grasses and herbs, whereas a newly sown ley often lacks herbs in the sward. Old turf also provides the thick sward which is important to maintain the soundness of young growing stock and will stand up to poaching better than newer leys.
- ☐ Rich pasture makes most animals fat and more liable to laminitis; sudden access to lush growth can cause laminitis even in thin ponies. A strict system of controlled grazing may have to be practised.
- ☐ Wet, marshy fields. Animals do not thrive in this type of field in winter. In summer, as the grass is often of poor quality, it suits ponies likely to get over-fat.
- ☐ High, exposed fields lacking shelter are very cold in winter and provide no escape from sun and flies in summer.
- ☐ Very steep fields are only suitable for small ponies. Larger ponies and horses – particularly young stock – may develop back, hock and/or stifle strain.
- ☐ Small paddocks can be invaluable in providing limited grazing for small ponies, or a daily exercise area for a stabled horse. They can, however, become heavily poached in winter. In general, horses are happier and thrive better in larger fields where there is more room.
- ☐ Low-lying fields or those with a clay-based soil are not suitable for winter grazing. The land becomes poached – i.e. cut up, with the roots of the grass exposed and often killed by frost. The quality of the pasture then deteriorates. Animals standing in mud are reluctant to lie down to rest. Much of the hay fed in the field is trampled in and lost.

WATER SUPPLY (see also *Watering*, Book 7)

A regular water supply is essential for health. If piped water or a suitable unpolluted stream is not available, alternative arrangements have to be made. When grass is plentiful and rainfall normal, a horse may well drink less than 4.5 litres (1 gallon) a day; the extra water intake being provided in the grass. In hot weather, when the grass is dry, or in freezing weather when the horse's diet is hay or other dry food, his daily water consumption is much greater, up to 45-54 litres (10 to 12 gallons) a day.

Water Troughs

The best way to ensure a satisfactory supply of water is to pipe it to a carefully sited concrete or metal trough and control it with a covered ballcock system. As the horse drinks and the water level drops, the ballcock drops with it; this releases a valve and allows the water to flow. As the trough fills, the ballcock rises with the water level and the valve shuts off the supply. This system usually works without problems. However, on occasions, the valve can stick and restrict the water, or the ballcock can puncture, so

Metal water trough with a covered ballcock system.

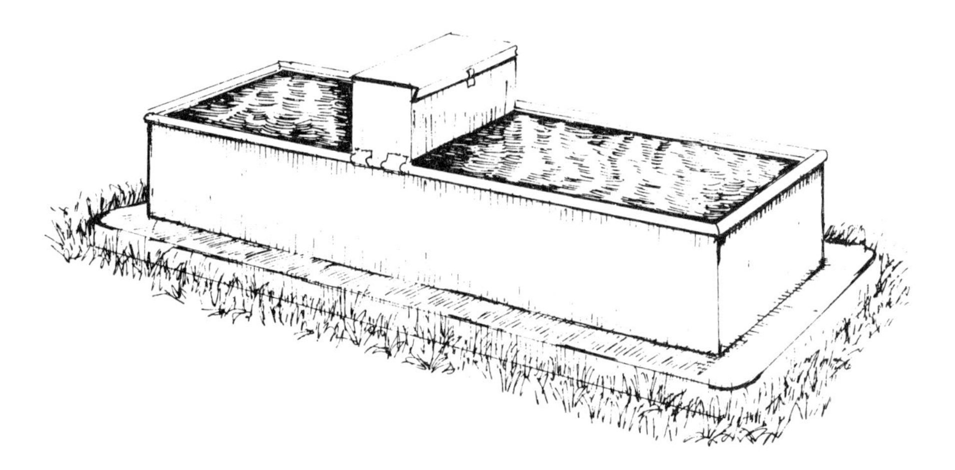

that the water will not be cut off and will spill over, causing flooding. Daily inspection is advisable.

Troughs vary in size and can hold as much as 909 litres (200 gallons). They should relate to the size of field and the normal stocking rate and should be large enough to allow several animals to drink at once.

Old baths and other metal and plastic containers are often used in small paddocks as a cheap alternative to purpose-built water troughs. These are suitable as long as there are no projecting edges likely to cause injury. Baths if boxed in are safer and less of an eyesore. If they are filled by piped water and a tap, the tap should be placed out of reach of the horses, so they cannot play with it or get caught up.

If they are close to a suitable supply, water troughs can be filled by means of a hose pipe. When the trough is full, the hose should be moved out of the way so that water will not syphon out. In freezing weather, the trough should be emptied. If water has to be transported, 22.5-45 litre (5-10 gallon) plastic water containers can be used. A strong wooden broom put through the hand grips enables two people to carry the container more easily.

Siting of a Water Trough

The cost of piping may influence the siting of your trough.

Boxed-in bath used as a water trough.

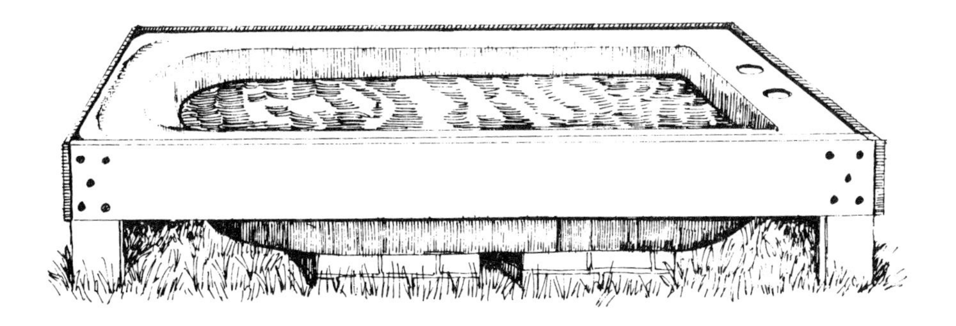

It should preferably be sited in a well-drained area of the field, away from the corner and near enough to the gate for easy checking. Do not put it near trees, as their roots make pipe-laying difficult and in autumn their leaves foul the water.

Troughs can be placed lengthways along a fence line, so that there is the minimum of projection on which a horse can knock his legs. In this position they can be made available to an adjoining field. Rails or boards built above the trough will prevent horses jumping out or fighting above it.

If sited away from the fence line, troughs should be at least 10 metres (33 ft) into the field and well away from the corner, so that there is less risk of a horse being trapped and kicked.

All troughs should be set on solid brick or concrete supports.

• The ground surrounding a water trough may become poached • Horses' legs can get very muddy, with the consequent risk of mud fever and cracked heels. • Thick mud may also discourage regular drinking. • To avoid these problems, loads of rubble, stone or rammed chalk can be put down, or a roughened concrete apron (approximately 10 square metres or 33 square feet) can be placed up to or around the trough (but note that in freezing weather the surface can become slippery, in which case straw should be put down).

Water troughs must be kept clean, so they should be regularly scrubbed out. During the process, the ballcock should be tied up. Troughs which do not have a base outlet must be emptied by bucket. A plastic dustpan is useful to finish the job.

Precautions in Freezing Weather

All exposed piping should be lagged and then boarded. The cavity should be filled with fibreglass. An alternative is to remove the bottom of an old dustbin or other container, put it over the pipe, and fill it with hot horse manure. It can be replaced as necessary. This system also helps to prevent the pipe freezing in the ground.

If metal pipes should freeze, they can be thawed out by covering them with straw and then setting light to it, or by using a blow lamp. Remember that as ballcocks are usually made of plastic they will puncture if exposed to flame.

With plastic piping, hot water can be poured over the pipe, which may also thaw the ballcock. Frozen plastic piping should not be subjected to direct flame: i.e. a blow lamp or burning straw.

When pipes freeze, they sometimes burst as they thaw. It is then necessary to turn off the water at the main. *The location of all field stop-cocks should be noted.*

In freezing weather, ice on a water trough must be broken each morning. Horses drinking during the day help to keep the water from freezing, but in very cold weather it may be necessary to re-break the ice at night. *These tasks must never be neglected.*

Rivers and Streams

This water supply can be satisfactory as long as the watering place is safe and the water unpolluted. There must be a sound approach area, sufficiently wide for several animals to be able to drink at once. If the stream is a field boundary, fencing should be constructed to prevent horses crossing over or moving up or down the stream. Streams with steep banks must have the approach area set well back so that the slope is less steep. Streams with a sandy bottom are not suitable, as a horse when drinking absorbs a certain amount of sand and this in time can cause sand colic. Streams adjoining or near buildings are often polluted by farm or stable drains, or by a neighbouring factory. If there is any doubt, a sample of water should be sent for testing.

Stagnant ponds should be fenced off, as they are not suitable drinking places.

POISONOUS PLANTS

Ragwort, buttercup, bracken, cowbane, deadly nightshade, foxglove, horsetails, meadow saffron, water dropwort and the leaves of most evergreen plants are all poisonous to horses. Fortunately, their bitter flavour ensures that as long as other grazing is available, they are rarely eaten. Occasionally an animal develops a taste for one of them and will suffer 'colic' or may even die.

After the use of weed-killer many of the plants listed above become palatable, so stock should not be allowed into a field which has been sprayed until all such plants have rotted down.

When found in hay poisonous plants are particularly dangerous – except buttercup, which when cut and dried loses its toxic properties. Old pasture which is about to be cut for hay should be closely examined, especially for ragwort.

Ragwort. Of the above plants, ragwort is the most common, and the one most likely to be eaten. In winter and early spring, young plants can be recognised by their prostrate growth and distinctive leaf shape. They bloom in July, August and September, when the yellow flowers can be easily seen. Mature plants grow to a height of 0.91-1.2m (3-4ft). Young plants may only be 10-15cms (4-6ins) in height. When cut, pulled, or poisoned by herbicides, they become palatable. If you are top-cutting fields in July and August, take particular care either to pull up the plants or to pick up the cut plants and burn them. This is even more important if the grass is cut for hay. The poison from ragwort acts on the liver and has a cumulative and long-term effect which is usually fatal. Grazing with sheep in the early spring will eliminate ragwort.

Yew. All parts of a yew tree are poisonous, even when the tree is dead. It appears to be palatable. Twigs falling off a tree are easily eaten with a mouthful of grass. Even a small quantity can be lethal. If a horse snatches at yew tree branches when being ridden through woods, the rider should dismount and remove all of it from the animal's mouth.

Deadly Nightshade. The berries, although rarely eaten by horses, are poisonous. They are brown to purple in colour. The plant is found as a creeper on hedges only in very limited areas. The less toxic plant, woody nightshade, which has red berries and a purple flower, is more common.
Hedges should be checked and any plants pulled out and burned.

Water Dropwort is only likely to be a problem after deep ditching.

Ornamental hedges and trees such as laburnum, laurel, privet and rhododendron, are more often found in the hedges or boundaries of fields adjoining parks and gardens. It is seldom possible to have them removed, so the area surrounding them should be fenced off. Laburnum seeds are particularly lethal. However horses are unlikely to eat them unless very hungry.

Acorns and Crab Apples. Fields containing oak and crab apple trees should be carefully checked in the autumn when the ripe acorns and fruit have fallen. Horses eat both acorns and crab apples. A small quantity does no harm, but a large quantity of acorns causes poisoning, and of crab apples severe colic.

Weeds such as daisies, docks, chickweed and broad-leaved plantain are of no food value. Their presence in a pasture reveals poor management. They are rarely eaten, unless cut with the hay. A large quantity of docks can then cause a digestive upset.

Nettles, thistles and dandelions are all relished by the horse when cut, but their presence signifies poor grass management.

LITTER

Fields which have roadside hedges or which are crossed by a public footpath should be regularly inspected for tins, broken bottles and other litter. In some areas, such fields require daily checking.

Fields in which rabbits are known to live should be regularly checked for holes. This is particularly important if foals or young stock are grazing them. Unsafe areas may have to be fenced off.

Fields newly acquired should be checked for abandoned farm implements such as harrows, which can be covered with grass and go unnoticed. Binder twine from used hay bales, or on occasions from rotted-down old bales, can also be very dangerous and should be picked up as it can entangle an animal's legs.

SHELTER

This is essential to protect horses from cold winds, driving snow and rain and in summer from hot sun and flies. A field bounded by thick hedges, stone walls or banks provides shelter in winter. Any trees give shelter from the sun.

Field Shelter

The only effective escape from insects is a field shelter. This is more likely to be used as a refuge from flies in summer than from cold in winter. If well positioned, artificial wind breaks made from stone sleeper walls or heavy plastic mesh, give welcome shelter in winter.

A field shelter should be placed either against the fence or well away from it, with the back of the shed to the prevailing wind. If placed well out in the field it gives some shelter from several directions. Open-fronted L-shaped sheds are the most practical, and they ensure that no horse can get trapped inside by another horse.

A well-positioned field shelter.

In poor draining areas, a concrete floor may be laid, with an apron extending well beyond the shed. A cheaper alternative is to put down loads of stone, rubble or rammed chalk according to availability. The floor area should always be kept well strawed down. Concrete can be slippery in freezing weather. On good draining soil, with the shed on a slight slope, a base may not be necessary.

Shelter sizes vary from a pony-type structure 3m x 3m (10ft x 10ft) to buildings of 12m x 18m (40ft x 60ft) which hold a number of animals. They must be strongly constructed to withstand horses rubbing against them. Some field shelter manufacturers insist on a concrete base. In certain districts, planning permission is necessary.

FENCES

It is the responsibility of the stock keeper to ensure that his fields are securely fenced. Animals which get out on a road can cause accidents, damage to life and property, and often fatal injury to themselves. They can also damage crops and gardens. It is possible to insure against these risks, but insurance companies rightly expect suitable care to be taken. An habitual escaper who causes damage may be responsible for his owner being taken to court. Subsequent insurance may be difficult to obtain and expensive. Animals which make a habit of getting out can rarely be contained against their will. A 1.5-1.8m (5-6ft) gate or fence is no deterrent, however electric fencing will usually contain them. (See *Electric Fencing*, page 38).

Small ponies can be successfully tethered or hobbled. Both these practices are open to abuse and should only be resorted to if all else fails, and only when a pony is on its own and not in a group (see *Tethering Ponies*, page 51).

If cattle or sheep are kept, either to run with horses or to provide alternate grazing, fencing must be strong enough to contain 'gadding cattle' (when pestered by flies in summer), and with the rails sufficiently close to contain sheep.

Owners of valuable stud and competition horses should take greater care with the choice of fencing. It must be secure and safe. If the animals are kept in adjoining fields it is best to separate them with a double line of fencing approximately 1.8m (6ft) apart, which prevents squealing and playing over the fence. An electric fence alongside the permanent fence will safely separate animals.

Hedges

A well-trimmed, thick hedge is the best type of field fencing. It should be kept at a height of 1.2-1.5m (4-5ft) and should be cut back annually to encourage growth at the bottom. This work can be carried out by arrangement with a neighbouring farmer or contractor. If it is done regularly by mechanical means the hedge trimmings will be chopped up by the machine and can be left to rot. Older, overgrown hedges have to be cut by a large circular-saw type of machine, and the large hedge trimmings should be collected and burned. Weak, long and straggly parts of a thorn hedge can be cut and laid. *Elder* should be dug out or cut down, as it weakens the other parts of the hedge-line; its only use is as a protection from flies, and for this purpose two or three large bushes can be left. *Holly* should be encouraged, as it gives good winter shelter. *Yew*, because it is poisonous, should be completely cut out.

Hedges under trees are often weak. Any gaps or weak parts should be filled with solid, creosoted posts and rails. Plain wire is inadvisable. Barbed wire should not be used. Horses standing under trees for shelter in summer stamp at flies and may put their feet over or through a wire fence.

It is natural for horses to browse. They may well take great pleasure in 'barking' any trees left unprotected. Plastic mesh placed round the trunks, or a protective guard of well-creosoted posts and rails, will keep horses away from trees: but note that link fencing or wire netting can be dangerous if horses strike or stamp at flies. Regular coats of creosote or tar will also act as a deterrent.

Walls

In certain parts of the country, stone walls are used as field boundaries. If they are of a suitable height and are kept in good repair they are stock-proof and safe, and provide good shelter. A low wall may need a rail above it to prevent animals jumping out. In the case of cobs and ponies, a cheaper alternative is one or two strands of well-tightened, plain wire or an electric fence.

Banks

Banks are found in Ireland, the West of England, and in some other areas. They give good shelter, but to be stock-proof low banks require topping with a hedge, rails or wire. Some banks have a deep drainage ditch on one side, which can be a danger to young stock and should be fenced off. Steep banks can present a hazard, particularly to foaling mares and young stock, and it is often safer to fence them off. It may also be necessary to fence across a ditch or stream at the field boundary.

Post and rail fencing.

Post and Rail
If in good repair, these make safe, stock-proof fences. Though expensive, they have a life expectancy of over twenty years, but can be damaged by chewing, rubbing and leaning. They should be regularly creosoted or painted • Broken rails must be promptly replaced as the split timber and any exposed nails can be a danger; morticed rails, though less easy to replace, are safer, as no nails are used • Fence posts should be sawn off at an angle, flush with the top rail. If horses try to jump out, projecting posts can cause damage.

Post and rail fencing is very expensive to erect.

Stud Rails
These make a safe, secure fence: they are more durable than wood, will not split, will not be chewed, and require less upkeep. In many cases they are also cheaper. From an environmental aspect, however, some people consider them to be less attractive.

Stud rails consist of plastic strips 10cms (4ins) wide, incorporating high-tensile wire. The strips come in rolls and are erected either on wooden or plastic posts. Three or more lines of rail strips are used.

A cheaper but less substantial alternative is a plastic

Stud rails with a plastic rail at the top.

rail at the top with two or three lines of plain wire below. This fence requires very exact straining. It stands up to considerable impact and also has some resilience. In time it stretches and requires tightening.

High-Tensile Wire Mesh Fencing

This is a heavy-duty wire mesh, preferably topped with a wooden or plastic rail to give clear visibility. It is cheaper than stud rail, and is escape-proof against any stock, but it presents the risk that animals – particularly foals or small ponies – may put their feet through the mesh and get caught. Small, diamond-shaped mesh will considerably lessen the risk, but horse owners should be aware of dangers. Mesh fencing must be well maintained and checked for any holes and tears.

Plain Wire

If well erected and firmly strained, this can be the economic answer to many fencing problems. Topped with a rail, it makes a more solid, clearly-sighted fence without too much expense. The drawback is that cobs and ponies may loosen the wire by putting their heads through and leaning on it; this can be prevented by using electrified wire or a line of barbed wire. High-tensile steel wire is more expensive than ordinary wire; it is less easy to work with and requires greater skill to erect, but it lasts twice as long. However if horses cut themselves on it the wounds tend to be very severe.

Barbed Wire

Barbed wire is not recommended but it is cheap and effective, and if it is well maintained and regularly inspected the risk is diminished. If barbed wire must be used, the safest place to put it is on the second line of wire with a single strand of plain wire above and two strands of plain wire below.

When renewing a fence, all old wire must be taken up and removed. To reduce the danger of a horse or pony getting a foot over the wire, the bottom strand should be of plain wire and never be lower than 28cms (15ins) from the ground. Another hazard is that of animals in adjoining fields squealing and stamping at each other through the fence; this can be dealt with only by careful daily observation.

ERECTION OF FENCES

Post and Rail and Wire

Posts should be at a minimum height of 1.07m (3ft 6ins), preferably 1.2m (4ft) above the ground and 46cms (18ins) below the ground. Posts may be of timber or concrete. All timber should be treated with preservative. Droppers (lightweight wood or steel slats) can be used between posts to keep wire taut.

Heavy strainer posts should be placed at all corners, at changes of direction, and/or at intervals of 50m (54yds). They should be dug 0.76m (2ft 6ins) into the ground. Stud rails and stud fences can usually be erected by the suppliers. For further information, consult the Ministry of Agriculture – Pamphlet No 711 *Wire Fences for Farms.*

Electric Fencing

This is cheap and easily erected. Its main uses are:

Electric wire using wooden posts.

- ☐ To keep horses away from a weakened hedge-line or fence.
- ☐ To provide safe separation from horses in adjoining fields.
- ☐ To divide fields temporarily so that areas can be rested.
- ☐ To permit strip-grazing of a field.
- ☐ To help prevent horses/ponies jumping out of their field.

It should be placed 1.5-1.8m (5-6ft) from the boundary fence.

The fencing usually runs off a portable unit containing a 6-volt battery. However, if it is close to a mains electricity supply, it can be wired to this, with a transformer which will reduce the voltage to approximately 6 volts.

The fencing material consists of either:

- ☐ Iron or plastic-coated posts with insulators at the top. They can be heeled into the ground. The latter are less likely to cause injury.

OR

- ☐ Wooden posts with insulators attached. This type is preferred for permanent fencing. Iron posts are easier to erect if the fence is temporary, or has to be moved each day as in strip-grazing.
- ☐ Plain round wire or wire strips which is stretched

between the posts and attached to insulators at either end. The latter can be more easily seen.
It is essential to cut back any long grass or overhanging hedge so that there is no risk of earthing the current, which – particularly in wet weather – will very quickly run down the battery. It is important to check the wire each day and to test the strength of the current with a fence tester. Horses are quick to take advantage of a weak or dead battery.

When erecting a fence, sharp corners and angles should be avoided, unless the posts can be efficiently stayed • The fence should not be placed underneath or parallel to overhead electric cables, as there can be some risk of the current being diverted down the wire • When moving a fence it is important to ensure that all posts are well heeled in, that insulators are sound, and that attachments to hedges and gate posts have an insulator to break the current.

Plain wire, preferably of the breakable type, can be more easily seen by animals if pieces of coloured cloth are tied between posts.

When first introducing animals to an electric fence, it helps to make them respect it if they are given a shock by feeding them close to the wire. Some horses never gain that respect, and happily jump out. Ponies with thick manes often run underneath. Such animals can sometimes be controlled by putting up two parallel lines of electric fencing. If this fails, they should not be turned out in a field with electric fencing, as they will encourage others in the field to imitate their bad habits.

Unsuitable Fencing
For safety reasons, the following fencing cannot be recommended, as there can be a risk of horses getting caught up:

☐ Wire netting.
☐ Sheep or pig wire.
☐ Chestnut fencing.

- ☐ Old iron rails. These are likely to spring out of line, and the iron spikes can cause nasty injuries to horses.

GATES

Wooden Gates

Oak is the most suitable type of wood. It is expensive, but it does not warp, and it lasts for forty years or more. Soft wood gates, though much cheaper, are inclined to warp, and the timber breaks more easily.

Gate-posts should be of oak, set 0.91m (3ft) in the ground. They should be hard-rammed with stone filling, or set in concrete. Old railway sleepers make a suitable alternative.

All timber should be treated with preservative. Hinges should be of galvanised steel and of a strength suitable to the gate.

Fastenings should be of a secure type, with no sharp projecting edges on which a horse can get caught. When fastened, they should help to take the weight of the gate, preventing it from dropping.

Metal Gates

If of heavy-duty construction, metal gates are satisfactory. Lightweight metal has a limited life; it is easily bent and the spars can fracture or come apart from the framework, exposing dangerous edges. Posts are usually of metal set in concrete, but wooden posts can be used. All metal work should be treated with rust-resistant solution and kept regularly painted. The hinges and fastenings used for wooden gates are suitable.

Hanging Gates

If gates are to last, they must be well hung, so that when opened they clear the ground and do not have to be lifted or dragged.

Padlocks

Owing to the increase of horse thefts, it is advisable to

padlock any gates opening on to roads. The hinge end of the gate should be secured by a chain and padlock, or by a metal bolt driven in above the top hinge so that it cannot be opened by lifting it off its hinges.

Gateways

Gateways used in winter become very muddy. They can be made up with loads of stone, chalk, or rubble, or with a roughened concrete apron. Surface water should be directed off the surrounding area by drains or ditches.

FEED

Summer

In summer, many horses and ponies at grass become too fat. If they are resting or only in light work (four hours a week), they are unlikely to require extra food. If they are on good grass, their grazing time may need to be restricted. If they are in heavier work and if the grazing is poor or limited, they may need a daily or twice-daily ration of hard feed. For amounts to feed see *Feeding*, Book 7.

Winter

In winter, supplementary feeding is essential. Hay should start being fed as soon as the horse shows interest when it is put out in the field • As long as there is grass to eat, they prefer this to hay • At first, they may only require hay at night, but as the grass diminishes, hay should be fed in the morning as well • If good hay is fed and not cleared up, this means that too much is being fed, and the ration should be cut down until all is cleared up • If the hay is of inferior quality, it will often be left. This should *not* be taken as a sign that too much is being put out, but rather that the quality of the hay needs improving • In wet weather, some hay is bound to get spoiled and to be trodden into the mud • In frosty weather, when the ground is hard, the horses should clear up all hay put down.

Methods of Feeding Hay

If one or two animals are kept, hay may be fed in haynets, which must be securely tied to a fence or tree and placed at least 3.6m (12ft) apart • *In wet weather* all animals prefer to eat with their heads down and their backs to the wind, so haynets may then be unwelcome • If more than two animals are turned out together, hay should be fed on the ground. It should be placed in the most sheltered part and on the least muddy area, even if this involves taking it to the far end of the field • Hay should be put out in slices, according to weight, and not shaken up as it will then be less likely to be spoiled • Heaps should be at least 3.6m (12ft) apart, and there should be two or three more heaps than horses to allow for them moving about • If there are ten or more animals and if the weather permits, it is better to put the hay cut in a larger circle. To check any fighting or bullying, wait and watch until all horses are settled and eating.

Quantities

Good-quality meadow hay is suitable for feeding in the field. Best-quality seed hay is a needless extravagance. If there is little or no grass, the following feed scale is suggested:

Small ponies	4.5-5.4kg (10-12lbs) hay
Medium ponies	4.5-6.8kg (12-15lbs) hay
Large ponies and cobs	6.8-9.0kg (15-20lbs) hay
Large cobs and horses	11.3-13.6kg (25-30lbs) hay

When the horses and ponies are of mixed sizes the hay ration at the beginning of winter should be about 4.5kg (10lbs) per animal. As winter progresses, this ration can be increased to 6.8-9.0kg (15-20lbs) per animal, depending on the weather and if it is being cleared.

In snow or freezing weather, extra hay *must* be given, it will not be wasted. Animals require extra food for warmth, and the process of eating and digestion helps to keep them warm and content. Their daily ration cannot be equated with that of a similar-sized horse kept in a warm stable. More food is necessary.

If hay supplies are short, oat or barley straw can be fed on its own, or the straw can be fed mixed with hay. Well-bred animals will then require a high protein supplement.

Concentrates (See also *Feeding* Book 7)
Healthy, resting ponies or those in light work (two to four hours per week) do not necessarily need a daily feed • Ponies under 12.2hh rarely require extra feed • Ponies and cobs in heavier work require a twice-daily ration of hard food suitable to their size • Horses wintered out should receive a twice-daily feed • All animals in light or poor conditions should receive as much food as they will consume.

When feeding concentrates, if only two or three animals are kept they may be fed in the field, providing an attendant stays to prevent bullying and fighting. This is not safe with a large number, and they should be brought in and fed separately.

Feed Blocks
These are manufactured specifically for feeding to horses and ponies wintered out. Together with rough grazing and/or hay, they provide a balanced and nutritious diet • They are weather resistant and can be placed on the ground in a container or suspended from a fence post or tree. They should be placed well away from the water supply • For groups of horses, allow one block per four to five horses and follow the maker's instructions. Generally speaking, horses only take up as much as they need and there should be no problem. However, certain horses develop a craving and finish a block in a day or within days. This makes it an expensive feed, and in some cases it has been known to cause severe colic.

COMPANY

Horses are by nature herd animals and they are much happier if part of a group. In summer, standing head to tail they are able to ward off flies. If it is not possible to find another horse for company, then a donkey or even a goat can be a successful substitute.

To avoid fighting and possible injury, it is important that groups of horses turned out together should agree. It is often arranged that mares and geldings are kept in separate fields. For this reason, many riding schools only keep either mares or geldings. If groups are mixed, be sure that you do not put two aggressive leader-type animals in the same field. Generally speaking, once animals know each other and have *sufficient room* they sort themselves out and there should be no trouble. However, care must be taken when adding a new animal to an established group.

DAILY CARE

Horses should preferably be visited twice a day, and on one of these visits they should be checked at close quarters. An observant person quickly notices if all is not well, particularly if the same person checks the horses each day. Any unusual behaviour or stance will then be noticed. A horse who is standing away on its own, or who is unusually slow or unwilling to come up when food or hay is put out, should be more closely examined. In areas where horses have been vandalised, a daily check of the whole horse, especially the dock area, is advisable.

Horses should look well and healthy with:

- ☐ A bright, alert expression.
- ☐ Weight evenly taken on all four feet, or a hind leg only rested. Well-bred horses and young stock sometimes flex a knee when grazing. This is normal.
- ☐ No discharge from eyes or nose.

- ☐ No stiffness or unevenness of stride when moving off.
- ☐ In summer, the coat lying smoothly and showing a gloss. In winter the coat may be rough, but it should not be dull and staring. A horse with a tight 'staring' or standing-up coat is cold and may be ill. He should be caught, stabled and his temperature taken.

Resting horses or young stock should be caught up at least twice a week and their feet, eyes, nose, skin and coat checked.

When observing a group of horses, a tit-bit may be handed out from a pocket, but a feed or catching bowl encourages competitive aggression, with a risk of injury to all concerned.

The water trough and fencing should be checked every day.

NEW ZEALAND RUGS

Although supposedly designed to keep the horse warm and dry, very few of these rugs are really waterproof.

Two rugs are necessary for a horse living out all the time, so that one is on the horse and one is drying. Even in dry weather the rug should be removed daily, the back rubbed to restore circulation, and the shoulders and legs inspected for rubbing. Rugs with surcingles are only suitable for horses being turned out for short periods, as the surcingles shrink in the wet and also cause pressure on the spine.

HEALTH

All horses should be immunised against tetanus, and a regular worming programme should be carried out. A careful worming strategy is essential if horses are to remain healthy and the level of larvae on the pastures kept to a minimum. The tactics are to use a suitable (veterinary-advised) anthelmintic at correct intervals and according to the time of the year (see Book 2). Teeth should be checked.

In summer many horses and ponies become grossly fat, with a consequent risk of laminitis, and if they are ridden the extra weight puts strain on the legs, heart and lungs. In many cases it is necessary to restrict their diet. Unless the paddock is bare of grass, one or two hours' daily grazing is ample. The animals may be stabled at night, bedded on shavings, and turned out in the day; sun and flies will then often deter them from eating. A small amount of hay can be given at night: *a pony* 0.9-1.4kg (2-3lbs); *a cob* 1.8-2.7kg (4-6lbs). If a secure yard is available, it can be used instead of, or in conjunction with, a stable.

In winter a pony's thick coat can disguise poor condition. However, if the crest, backbone and rib areas are felt, any lack of flesh will be apparent. The crest in particular is a reserve of fat and it should feel firm and solid. Well cared for healthy ponies should look plump and well throughout the winter, whatever the weather. Ponies with staring coats and bony outlines are a sign of poor management, inadequate worming and insufficient food (see Book 2).

Seasonal Problems

In Summer

Laminitis. Ponies subject to laminitis must be kept on a strict diet. If there is any sign of 'footiness', heat in the foot, or standing with weight on the heels of the forefeet, take the pony off grass and consult your veterinary surgeon.

Sweet Itch. The signs are continual rubbing of the mane and tail until they become raw. Only certain horses – usually ponies – appear to react to the saliva of the evening midge which causes the allergy. The condition is improved if the horse is stabled before dusk and until well after dawn. Fly repellent should be used in the stable and in the field. It has been found that if the wind is over 3mph midges will not appear.

Blistered Muzzle. Horses with white muzzles are particularly prone to sun-blistered mouths and noses: a condition which can be caused by an allergy to a particular plant. To prevent this, apply barrier cream. If blistering occurs, antiseptic ointment should be used. It may be necessary to stable the animal.

Running Eyes. Flies settle round a horse's eye to feed on 'the tear'. Sensitive horses can wear a headcollar with a fly guard (browband with a string fringe). More sophisticated protectors, such as browbands impregnated with fly repellent, are now available from tack shops. Some fly repellents are safe to use round the eyes, and can be applied with a small sponge. Horses can be kept in during the day. The eyes should be sponged clean twice daily. If eyes become infected, veterinary advice may be necessary. Light-skinned horses are more subject to this problem.

Bot or Gadfly. From June to September the bot fly lays its eggs on a horse's front legs, particularly just below the knee. These hatch to larvae, which the horse licks. The larvae then penetrate the skin and the mucous membrane around and inside the lips, resulting in the start of the bot fly cycle. The flies cause great irritation, and although they do not land on the skin, horses will gallop about and kick to get away from them. They may then injure themselves and others.

Warble Fly. These flies are becoming much less common, but any swelling on the horse's back in spring and early summer should be suspect. When the larvae hatches out it must be killed. Some modern wormers kill bots and warble larvae before they reach their final sites.

Various fly repellents are now available, which in warm weather help to make horses more comfortable. Many repellents have to be applied twice daily, but some are longer acting and in dry weather are effective for several days. A paraffin rag wiped on the legs can be effective against bots, but should not be used on horses with sensitive skins, as it may cause blistering.

In Winter
The following problems are likely to occur, particularly with fine-coated horses in a very wet season. Horses should be closely observed each day and preventive measures taken. At any sign of lameness, obtain veterinary advice.

Rain Scald. The skin of the back, loins and/or quarters is softened by incessant rain, thus weakening the natural defence mechanism against skin germs. Inflammation and dermatitis follow; the hairs clump together and fall out in patches.
Prevention: Sensitive-skinned horses should wear New Zealand rugs. In autumn, liquid paraffin or barrier cream can be applied to the horses' backs when the coat is free from mud. In severe cases, consult your veterinary surgeon.

Mud Fever. This is similar to rain scald, but occurs on the legs and sometimes the heels. If on the latter, continual exposure to mud increases the infection. Severe cases may spread to the belly and inner thigh.
Prevention: In the autumn and during the winter months apply barrier cream regularly to the legs, after they have been cleaned and dried. Wet, muddy legs should never be brushed, but thatched and left to dry (see *Bandaging*, Book 4). Then they can be brushed clean.

Cracked Heels. These are a chronic form of mud fever. Cracks in the heel and pastern area become intermittently infected, which can cause acute lameness, particularly after the animal has been standing still.
Prevention: As for mud fever.

Thrush. This is a disease of the frog. Most cases are associated with poor foot trimming or even foot disease. However, some horses will contract a mild form of thrush in

winter through standing in mud.
Prevention and treatment: As soon as the ground becomes muddy, apply *Stockholm tar* in and around the frog area. Repeat as necessary. The foot and frog must first be thoroughly washed clean, and dried. The veterinary surgeon may recommend an antibiotic spray as treatment.

Lice. These insects can be found in the coat at any time in the winter, but usually in and after January. Any horse rubbing its mane and tail should be suspect.
Prevention: Apply louse powder regularly, as directed by the manufacturer. If this proves ineffective, obtain veterinary advice, as more sophisticated treatment may be necessary.

Ringworm and Other Fungal Infections. These can appear at any time of the year, but in winter the fungus in its early stages can go undetected, as the long coat covers the scabby area.

Horses at grass, or those brought in from grass to be stabled, should be closely examined for any raised area of hair or circular scabs. The fungus can appear anywhere on the body, but it is more likely to be in places where the horse has rubbed against a gate or fence post: i.e. head, shoulders and quarters.
Prevention: Particularly if cattle have been in the field, all gates and fences should have a thorough coating of creosote. Ironwork should be washed with a strong solution of soda or repainted.

Sudden Acute Lameness. This can usually be attributed to a punctured foot or a small flesh wound, which has gone septic. It may be necessary to transport the animal to the stables. The foot and leg should first be thoroughly cleaned and then examined for injury. Veterinary help should be obtained.

TETHERING HORSES AND PONIES

From many points of view, tethering as a method of managing horses and ponies is unsatisfactory. If it has to be used, the following 'Code of practice' is recommended:

Choosing the Site

- ☐ The area should be well grassed (especially if grass is going to be the sole source of food) and free from poisonous plants, shrubs, and trees: e.g. ragwort, laurel and yew.
- ☐ The site should be reasonably flat.
- ☐ An area of shade should be available. This is essential during summer.
- ☐ The area of tether must not cross a public footpath or be close to a public highway, especially where there is fast moving traffic.
- ☐ The ground must not be waterlogged.
- ☐ There must be no large objects within the area of tether, e.g. tree stumps or bushes around which the tether can get entangled.
- ☐ The size of the site must depend on:
  The amount of grass available.
  The frequency that the horse will be moved.
  The minimum size should be a 6m (20ft) radius.

Types of Tether

- ☐ A wide leather neck band fitted with a swivel. It should not be attached to a headcollar.
- ☐ A firm stake.
- ☐ A metal chain of appropriate weight for the horse to be tethered, attached at stake with a 360° swivel fitting at ground level, allowing the animal to cover a complete circle without entangling the tether.

CHAPTER 3

Working the Grass-kept Horse or Pony

Grass-kept horses and ponies can be used for hacking and light to medium work. Highly strung or 'gassy' animals may then be more amenable. Horses needed for hard work can be kept at grass in the summer, as long as they are given regular exercise and increased concentrates • If the horse starts to get too fat, grass intake on a good pasture may have to be restricted • In winter, if hard, fast work is required, the horse should be stabled at night and regularly exercised each day.

Summer

☐ Horses and ponies on good grazing should preferably be brought in two hours before being ridden, so that their stomachs are not gorged with grass.

☐ For grooming and care after work, see Book 2, *Care of the Horse*.

On a wet morning, the horse's back should be scraped with a sweat scraper and semi-dried with a stable rubber or straw. If a dry stable rubber or numnah is put on and secured by a surcingle or saddle, the back will quickly dry. The horse will not be harmed if ridden with a wet back, although this is not recommended: he will probably react by first hollowing under the weight of the rider; a sharp horse may even buck.

Winter

- ☐ If on hard food, the horse should be fed early in the field, or caught about 1½ hours before being ridden and given his feed.
- ☐ For grooming, see *Book 2*.

Wet and muddy backs should be scraped with a sweat scraper or plastic curry comb, dried as above, then brushed clean. All lumps of mud must be removed from the saddle and girth area, and a fresh numnah or stable rubber should be placed under the saddle.

If on returning from work the horse/pony is warm, any resulting sweat patches should be wiped clean or sponged off • Feet should be picked out and the animal should be turned out as soon as possible, so that he can roll and be less likely to catch a chill • If he is wearing a New Zealand rug, straw can be put under it to allow more air to circulate, and he can be given hay. His hard feed can be given in the field, or he can be caught later and fed in the stable.

If it is essential to keep the horse or pony in after work, put on either (a) a sweat rug and roller with straw underneath, or (b) a sweat rug and light blanket and roller • The horse/pony may then be fed • It is important that ponies with thick coats are not allowed to sweat profusely.

THE COMBINED SYSTEM

This is a system whereby in winter the horse is stabled at night and put out in the day. In summer, the process is reversed: the horse is in during the day and out at night, thus obtaining protection from hot sun and flies. If time and labour are short the advantages are:

- ☐ Daily exercise is not necessary.
- ☐ If only ridden at weekends, the horse will be quieter and more manageable.
- ☐ Daily grooming is reduced.
- ☐ Less bedding is used and at times less hay.

☐ Mucking-out time is reduced.
☐ The horse can be clipped and turned out in a New Zealand rug.
☐ The horse can be made ready for work quickly, as he will be dry and comparatively clean.
☐ In winter, less food is wasted in the field, and the food ration can be reduced, as the horse will be in a warm, dry stable at night.
☐ In summer, if necessary, grazing time can be more easily controlled.

YARDING PONIES

This is a system whereby ponies are wintered in a covered yard or large barn, which may or may not have an adjoining railed exercise area. It is a system very suitable for riding schools. Pony mares and foals can also be kept in this way.

The yard is kept strawed down on the deep litter system and is cleared out at the end of the winter • Hind shoes should be taken off and, if feasible, front shoes also • Mares and geldings may have to be separated, or only a single sex kept or accepted for winter keep • An area of 20m x 20m (65ft x 65ft) is sufficient for ten ponies • A hay and feed manger should run along two sides. • Preferably there should be two separate water troughs.

Advantages
☐ Ponies will be clean, dry and warm.
☐ Less time will be needed to groom before being ridden.
☐ They will be available when required.
☐ They will need less feed and there will be less waste. (See *The Combined System*, page 53.)
☐ If access is suitable, the yard can be cleared by a 'muck lorry'.

Disadvantages
☐ Certain ponies do not settle and may fight and upset the group.

☐ Fights may occur, resulting in injuries. At feed time, it is sensible for someone to be present to prevent squabbling.
☐ Labour and suitable equipment will be needed in the spring to clear the yard, unless access is suitable for a muck lorry.

CATCHING HORSES/PONIES

Always take either a tit-bit or a bowl of food to the field • Do not be in a hurry. Careless procedure can easily upset a horse, spoil him, and make him difficult to catch • The headcollar should be placed over the left arm. The bowl or tit-bit should be held in the left hand • The rope in the right hand, approach the horse from the front, and as he accepts the food, move towards his left shoulder, placing your right hand with the rope on and then over his neck • Avoid any sudden or jerky movements, and talk quietly to the horse throughout the procedure • Place the bowl on the ground and quietly put on the headcollar.

Procedure with Difficult Horses/Ponies

☐ Take a large feed bowl, half fill it with food of a type which will rattle when the bowl is shaken. Tuck the headcollar and rope under your left arm, or alternatively hold a rope under your arm or inside your coat.

Always approach the horse from the front • If he moves away, make a big circle and again approach from the front • Talk to him as you approach, trying to look him in the eyes • Shake the bowl, and then lower it to the ground, which may gain his attention • Walk up to him with the bowl in the left hand • As he takes the food, the right hand can quietly move up his jaw and towards his neck • If he moves away, do not try to hold him: try to go with him, but if he shoots off, do not follow. Reorganise your approach as from the start.

When he stands and eats, move quietly round to his

shoulder, taking hold of the rope with your right hand • Place the rope under and then up or over and down his neck • When the horse is held by the rope, you should be able to put on the headcollar • On some occasions, a length of string or thin rope may be less obvious.

If difficult horses are caught and fed each day, it becomes a routine, and when the animals are wanted for riding, they will be easier to catch.

Difficult horses are usually more easily caught if one person only makes the attempt and any onlookers are sent away. Other horses in the field should be caught and held at the gate or in the corner of the field. At first do not take them out of the field, as a horse left on his own will be encouraged to gallop about. However, if there is a convenient passageway or track to the stable yard, then the horse will probably follow the others along it, or he may walk in amongst the groups and allow himself to be caught. If this occurs, do not approach from the front, but from the rear. Talk to the horse and place a hand on his quarters (taking care he does not kick you), moving it up his back and neck as you walk alongside him. Secure him by placing a rope round his neck, and then quietly put on the headcollar. Horses must not be allowed to escape on to roads.

If the horse still will not be caught, leave a quiet horse or pony in the field with him and return in half an hour.

Horses are usually easier to catch in hot weather and more difficult in wet and/or windy weather, or when there is spring grass • The catcher can often put horses off by wearing an unusual hat or raincoat, particularly if it is plastic • A better contact is made with the horse if gloves are not worn.

Horses who are difficult to catch are best turned out wearing a well-fitting leather or nylon headcollar. If of nylon, it should have a 'panic break', which will give way if the horse catches the headcollar on a post, etc. A 16cm to a maximum of 20cm (6-8ins) length of knotted string can be attached to the centre 'D' and used when catching the animal; the string should never be in a loop. The horse's head should be regularly checked for sores and rubs.

Staff (e.g. an instructor or groom) catching a horse and riding it to the stables should first put a headcollar on the animal and then fit the bridle on top of it. If catching the horse with only a bridle, do not put the reins over first in case the horse pulls away and takes the bridle with him.

On no account should animals be chased about or rounded up. This is self-defeating, and can only make them upset and even less willing to be caught. A sufficient number of helpers (a minimum of six) can form a straight line and walk towards the horse, guiding him into a corner. Even then, if he is determined, he can easily outwit the plan.

Remember that the 'boss' is responsible for the safety of any helpers. Though horses and ponies are usually less upset by children and will accept them if they move near, fear of being caught may alter this attitude.

Unaccompanied small children should not be allowed to catch ponies, particularly if there are several ponies in the field • Only capable children should be allowed to ride either to or from the field, and then with great care • Hard hats *must* be worn • The horse must always be bridled.

TURNING OUT HORSES AND PONIES

It is important that safe procedure is followed. Horses will often kick out when first released, and can injure people or other animals. When leading a horse through a gateway you must be on the side nearest the gate, to control and stop it swinging back and hitting the horse.

Procedure for One Horse

☐ Bring the horse through the gate.
☐ Turn his head towards the swinging end of the gate.

☐ Shut the gate.
☐ Take the horse several yards into the field and once again turn his head towards the gate and yourself.
☐ Pat him, take off his headcollar, and step back.
☐ Watch the horse as he moves off. Do not turn away.
☐ Never slap him on the neck or quarters, as this may encourage him to gallop.
☐ Try to proceed quietly and avoid any fuss. If he is wearing a bridle, follow the methods listed above, then undo the throatlash and noseband (and the curb chain if you are using one).
☐ Take the reins over the horse's head and drape them around his neck.
☐ Hold the reins firmly under the neck in the right hand, and with the left hand ease the headpiece over the horse's head. This allows him to quietly release the bit without worry.
☐ Pat the horse on the neck and release the reins.

Procedure for Several Horses

☐ All horses should be brought into the field, taken well away from the gate, with at least 3m (12ft 9ins) between each horse, and turned to face the gate.
☐ Shut the gate and proceed as directed above.
☐ Finally, release all horses together.

Horses who are unfamiliar to each other or who are likely to gallop off, should be taken to the field in pairs. When the first pair have settled, the next pair can be taken out and released.

Roughing-Off a Fit Horse

This is the preparation of a fit horse for a holiday and a rest out at grass. The rest is important both from the physical point of view and to relieve mental stress.

Rests are given to:

☐ Hunters at the end of the season.
☐ Competition horses.

☐ Injured horses, to allow the repair processes time to work.

The roughing-off process takes about two weeks and the procedure is:

1. Remove one blanket. After seven to ten days remove the top rug.
2. Thorough grooming should be discontinued and the natural grease should be allowed to accumulate in the coat, forming a protection against cold and wet.
3. Diet should gradually be adjusted, reducing the hard feed and increasing the hay.
4. The horse should be allowed out daily in a small field or led out to graze.
5. Exercise should gradually be reduced and should now consist mainly of hacking out at the walk.
6. Veterinary checks should be made, and any recommended treatment should be completed.
7. Shoes should be removed and feet should be rasped to prevent the horn splitting. Horses with brittle feet who are to be turned out on flinty or very hard fields may need to have their front shoes left on. Alternatively, grass tips can be fixed (see *Shoeing*, Book 2).
   NB: Feet should be rasped or shoes should be removed every four to six weeks.
8. In cold weather, the horse will be more comfortable if a New Zealand rug is used. For fitting and procedure, see page 46.

Hunters are often turned out at the end of the season. This can be as early as the end of March, when they will need a New Zealand rug, a well-sheltered field, and hard feed and hay to help keep them warm.

Preparation for Turning Out

1. Check the field.
2. If you are turning out a horse on to rich pasture, he must be given short periods of grazing for several days beforehand.

3. Choose a mild day.
4. Make sure the horse has suitable company – preferably not a highly-strung galloping type of horse.
5. Do not give a breakfast feed. If the horse is hungry he is more likely to graze than to gallop.
6. Turn the horse out early in the day, so that he will have time to settle in the field before night. He should be watched until he is quiet, and checked again during the day.

CHAPTER 4

Bringing a Horse Up From Grass

It is better for the horse if the change is made gradually over a period of about ten days, as this will allow his microbial digestion time to adjust itself to the change of diet (see *Feeding*, Book 7).

Depending on the time of year and quality of grazing, some horses may be receiving hard feed and hay in the field. In this case, a sudden change of diet will present no problems.

Procedure

- ☐ Bring the horse into the stable for several hours a day. Give a small feed of 0.45-1.4kg (1-3lbs) of 'Horse & Pony' cubes or equivalent foodstuff or 0.9-1.4kg (2-3lbs) of grain, plus a 1.8-2.3kg (4-5lbs) net of dampened hay. Allergic horses should be given soaked hay (see *Feeding*, Book 7).
- ☐ If the horse is turned out on his own or with one or two others, it is possible to feed them in the field. Always stay to watch the feeding, so that there is no bullying or kicking. (There is no need to feed hay.) This practice is not recommended if there are *more* than two or three horses in the field, because of the risk of kicking.
- ☐ The horse may have to be brought in and kept stabled. In this case, try to arrange for him to be turned out in

a paddock for a few hours each day, or, alternatively, graze him in hand for twenty minutes or so.

Feed ration should be two or three 0.4kg (1lb) feeds of 'Horse & Pony' cubes or their equivalent, and plenty of dampened hay, about 9.1-11.3kg (20-25lbs). Do not feed any grain for a week; grain fed too soon can be the cause of filled legs. If you are planning to use grain, change over to it gradually.

Usually if a horse comes in after a summer at grass, weight will need to be lost. Gradually accustom him to a smaller total quantity of feed, but one which includes more concentrates. (See *Feeding*, Book 7). As more concentrates are fed, hay can be reduced. If the diet is restricted to encourage weight loss, it is essential to put the horse on shavings or a similar bed, and not straw. For amounts of feed, see *Feeding*, Book 7.

Recommended Treatments

Worming. A suitable dose should be administered. (See *Worming*, Book 2.)

Flu and Tetanus Inoculations. If due, these should be arranged. (See Book 2.)

Teeth. The teeth should be checked by the veterinary surgeon and rasped if necessary.

Feet. Arrangements should be made with the farrier to have the horse shod as soon as he is permanently stabled and starts exercise • Four weeks before he is due to come in, depending on the condition of his feet have him shod with light front shoes. This allows some growth of horn and makes it easier to shoe him for work. Valuable horses should never be turned out in company with hind shoes on, as a kick from a shod foot can cause severe injury.

Precautions
With careful feeding and regular walking exercise, the problem of filled legs should not arise. Coughs and colds easily develop during this period, so ensure a plentiful supply of fresh air and take care if the horse sweats when exercised. In the past, horses were physicked when first brought in, but today this is not recommended. See *Feeding*, Book 7.

Dangers From and Precautions Against Gaining Weight
Turning horses out for long periods in the spring/summer may result in their putting on too much weight. The increase in weight puts strain on a horse's legs and must be reduced before he is fit for active work. Often the time allowed for fittening is insufficient, and horses are asked to work too early, which strains the tendons, joints, heart and wind, with the possibility of permanent harm.

Horses can be turned out for a shorter period of six to eight weeks. They can then be either yarded or stabled at night. In very hot weather, when there are many flies, they can be brought in during the day and turned out at night.

Suggested Exercise for an Unfit, Overweight Horse After Several Months at Grass

1st Week
Walking exercise on the level. Half an hour on the first day, increasing to one hour by the end of the first week.

2nd Week
Increase walking up to 1½ hours by the end of the week.

3rd Week
Slow trotting, starting on the level, together with walking up and down hills.

4th Week
As third week.

5th Week

Two hours' exercise a day, to include some trotting up hills and short, slow cantering on level, good ground. Sweating will help to hasten weight loss. Twenty minutes' school work may be included.

6th to 8th Week

Two hours' exercise a day, to include trotting up hills and longer cantering periods. School work may include jumping and canter circles.

7th Week

If the ground is hard, cantering can be performed in a school. Start with one minute on either rein, building up to not more than five minutes on either rein. As this can be very strenuous, care should be taken. Big, heavy horses unused to circle work should not be asked to do this type of work.

For the school, work boots should be worn all round. If, when hacking out, roads should be slippery, or if the horse is fresh, knee caps and boots are advisable.

All exercise periods should include a good percentage of walking, as this helps to harden legs and to develop muscle without unnecessary strain.

Horses who are turned out for shorter periods – or in the winter, when grass is poor, who have received regular hard feed and hay – come in very much fitter. In these cases, preparation for active work may be shorter. The owner or stable manager must decide how fit a horse is and for what work he is ready.

Unfit overweight horses should not be lunged, because working on a circle puts extra strain on muscles, tendons and joints. Lungeing may be included after two weeks of walking, but five minutes on either rein will be sufficient at first. This should have a beneficial and suppling effect on the horse, as long as a suitably prepared surface is used. Hard or slippery fields are not advisable.

Timetable for Fittening
If the horse is grossly fat the time taken over fittening must be increased.

Riding Club – after 6-8 weeks

Hunter – can be brought up in August ready for gentle cub hunting by mid-September; more serious cub hunting in October; and hunting by November.

Novice eventer – 10-12 weeks.

Advanced show jumper – 10-14 weeks

Long-distance horse 40-mile ride – 8-12 weeks

Dressage – 10-12 weeks

3-day eventer – 14-16 weeks

Golden Horse Shoe horse – 16-20 weeks

Prevention of Sore Backs, Girth Galls and Sore Mouths
☐ Check the fit of the saddle.
☐ Ensure that the bit fits the horse's mouth.
☐ All leather work must be soft and supple.
☐ Tack should be cleaned each day.
☐ Use a numnah under the saddle. Natural cotton or sheepskin are preferable to man-made fibres. In hot weather, a padded cotton numnah helps to absorb sweat without overheating the saddle area.
☐ Use a lampwick girth.
☐ Numnahs and girths should be regularly washed – if necessary, each day.
☐ Harden saddle and girth areas with salt and water or methylated spirit.
☐ The horse must be thoroughly cleaned each day; sweat marks quickly become sore if left • In warm weather, the horse should be sponged down; surplus water should be removed with a sweat scraper; then the horse should be walked in the sun until dry • In cold weather, sweat marks can be sponged off with warm water • In both cases the horse should be dried,

and thoroughly groomed • During grooming he should be checked for any signs of saddle sores or girth galls. The first signs are rubbed hair, followed by tenderness and heat.

Treatment

☐ Harden the area with salt and water. Stop riding, and either lead the horse out in a cavesson or bridle alongside another horse, or, if he is fit enough, lunge him • If the saddle sore is damp (scaled back), spread cold kaolin over the area and top it with a plastic covering. This will help to draw out the inflammation before hardening the skin with salt and water.

☐ Legs should be carefully examined for any signs of heat. Hard ground and flints can cause problems with bruised soles • It is preferable for horses with thin soles and/or flat feet to be shod with pads or leather soles.

Bibliography

ARCHER, MARYTAVY, *Pasture for Horses* (Riding School and Stable Management Ltd).
BRITISH HORSE SOCIETY, *Manual of Horsemanship* (Threshold Books).
BRITISH HORSE SOCIETY, *Keeping a Pony at Grass* (Threshold Books).
BRITISH HORSE SOCIETY, *Grassland Management fcr Horse and Pony Owners*.
SCOTT, CLIVE, *Grassland Management* (Private issue).
Management of Horse Paddocks (Horse Race Betting Levy Board). A.B.R.S. Pamphlet.

Index